The Deep Blue Planet

ALONG
THE
COASTS

The Deep Blue Planet

ALONG THE COASTS

Renato Massa
English translation by Neil Frazer Davenport

RSVP
RAINTREE
STECK-VAUGHN
PUBLISHERS
The Steck-Vaughn Company

Austin, Texas

Published by Raintree Steck-Vaughn Publishers, an imprint of Steck-Vaughn Company

Editors
Caterina Longanesi, Linda Zierdt-Warshaw, William P. Mara

Design and layout
Jaca Book Design Office

Library of Congress Cataloging-in-Publication Data

Massa, Renato.
    [Tra mare e terra. English]
    Along the coasts / Renato Massa ; English translation by Neil Frazer Davenport.
      p.   cm. — (The deep blue planet)
    Includes bibliographical references and index.
    Summary: Describes the characteristics of the two different types of coastlines that exist throughout the world and the plants and animals that are found there.
    ISBN 0-8172-4654-1
    1. Coastal ecology — Juvenile literature.   2. Coastal animals — Juvenile literature.
[1. Coastal ecology. 2. Ecology.] I. Title II. Series
QH541.5.C65M3813   1998
577.5'1 — dc21   97-12103
                                      CIP   AC

Printed and bound in the United States
1 2 3 4 5 6 7 8 9 0 WO 01 00 99 98 97

CONTENTS

Please note: words in **bold** can also be found in the glossary. They are bold only the first time they appear in the main body of the text.

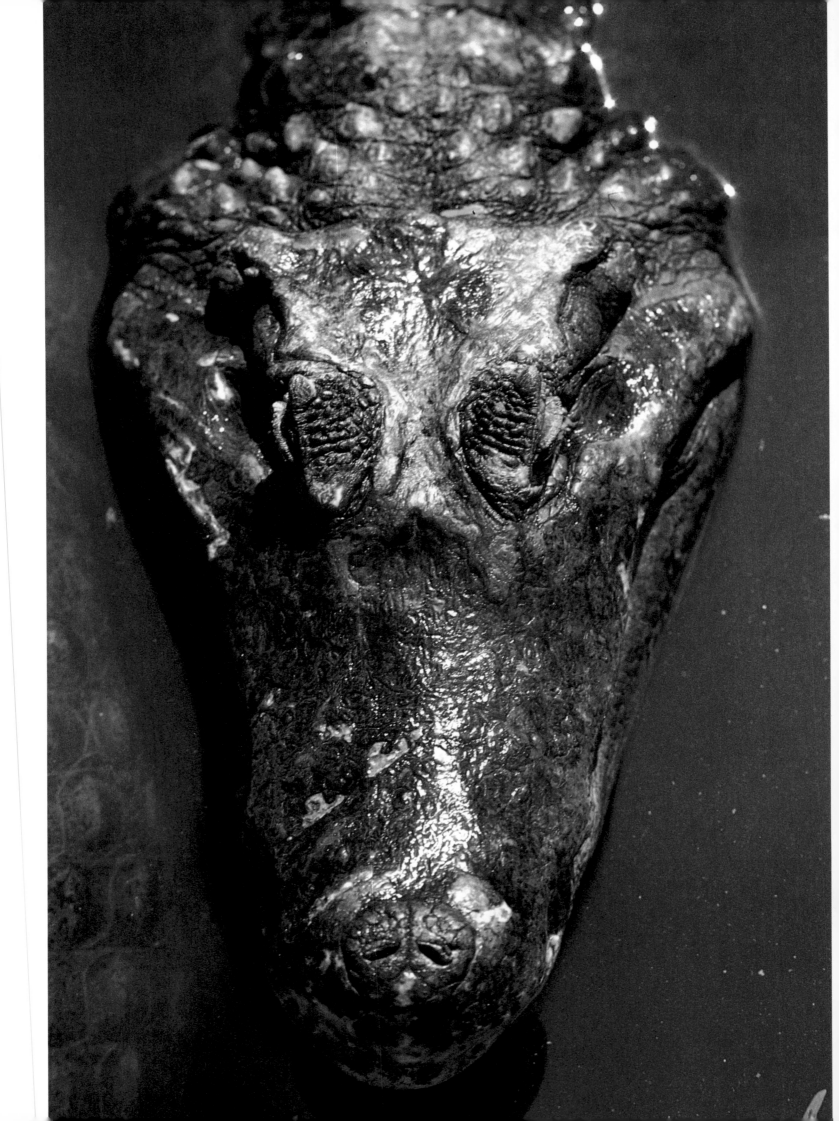

INTRODUCTION

The oceans and seas are wondrous places of both dark and light, flora and fauna, warm and cold, and life and death. Many regions are still largely unknown to us, so there is little doubt that hundreds of mysteries still lie in their murky depths.

The oceans and seas cover just over 70 percent of the Earth. They have provided us with many pieces to the puzzle of how life began on this planet, since the earliest forms of life were thought to have existed there. The rocks and sediment along the ocean floor have yielded a great deal of fossil evidence over the years.

The value of the oceans and seas is immeasurable. They have given us many useful chemicals and minerals, including bromine, magnesium, and salt, not to mention pearls for jewelry and shells for building material and health supplements. Most experts believe we have not yet realized their full potential in regard to nutrition, though it is believed that humans derive at least 10% of their overall protein from the Earth's waters, either directly or indirectly. Finally, there is the recreational aspect. Activities such as swimming, fishing, boating, diving, and so on, when executed properly and responsibly, provide us with a great deal of pleasure and a measure of relief from the grind of our daily lives.

Sadly, however, we humans have caused some serious damage to the oceans and seas in recent times. Industry is the greatest violator, with over a quarter of a million manufacturing facilities using the great bodies of water as dumping grounds for their often highly dangerous waste products, including mercury, lead, sulfuric acid, and asbestos. In addition, towns and cities regularly dump improperly treated sewage and millions of tons of paper and plastic wastes into rivers, streams, and lakes. Plastics in particular have the potential to remain intact for hundreds of years.

However, we have not yet reached a point of no return, and one of the goals of the *Deep Blue Planet* books is to give you a deeper understanding of—and in turn a deeper appreciation and respect for—the aquatic environments of this world. The more you know about any subject, the greater your appreciation for it will be, and the oceans and seas are in desperate need of increased appreciation. Perhaps someday you will make efforts of your own to preserve these beautiful natural areas and the myriad life forms that thrive within them. If so, you will be helping to guarantee them the bright and vibrant future they so richly deserve.

THE COASTAL FRINGE

1. A sandy beach on the Thailand peninsula. In the case of low coasts like this one, deposition is dominant over erosion. **2.** A high coast, Bempton Cliffs in northeast England. The sharply cut rocks plunge to the sea with a sheer drop of about 100 meters (328 ft). Seabirds nest on the cliffs.

A Frontier World

In many areas of the Earth, the sea meets the land to form a frontier called the coast. The coast may be the only part of the world in which people can clearly see the forces of nature at work. In these areas, the sandy or muddy sea floors, the rocks, the sandy beaches, the saltwater lagoons, the **deltas**, and the **estuaries** teem with life. This is the home of land organisms that have adapted to the sea, and **marine** organisms that have adapted to the land.

Waves, **tides**, currents, rivers, and storms regularly shift the borderline between land and sea. The coast becomes a buffer zone that is open to many creatures. The only

that rise only a few feet above the water level, or by towering cliffs that are completely inaccessible.

Sandy and muddy coasts are areas in which the **deposition** of **sediment**, brought to these areas by water, wind, and ice, plays an important role. Much of this sediment is made up of the skeletons of marine organisms. The mechanical action of waves and currents deposits particles in different areas. The particles are called gravel, sand, or silt, based on their size.

rhythm of their lives to that of the tides. They must be active and find food during high tide, and rest at low tide. During low tide, limpets, barnacles, and chitons will attach themselves to the rocks, retaining a little water and oxygen in their shells. Sea anemones will withdraw their tentacles into their gastrovascular cavities. Marine worms and many mollusks will hide in deep tunnels.

requirement for living in this zone is the ability to adapt to the constant changes.

Rocky and Sandy Coasts

For the most part, there are two types of coasts—rocky and sandy. Rocky coasts form when **erosion** by the sea is the dominant force of nature. The action of waves, tides, and currents gradually wears away "weaker" elements, such as soil and plants, leaving only bare rock. The rock may then be broken down according to its chemical and geological nature. The sea may be bordered by small piles of rocks

The Intertidal Zone

The area between the high tide and low tide marks is called the intertidal zone. The intertidal zone is subjected to the daily invasion and withdrawal of water. This area may be only a few dozen centimeters wide, or many meters. It depends on the location.

All marine organisms that make their home in the intertidal zone must adjust the

The intertidal zone is searched, at intervals of about six hours, by **terrestrial** and marine predators. During low tide, the area is invaded by birds searching for the opportunity to extract mollusks hidden in the sand and silt. As high tide returns, the fishes will exploit the remains of the previous cycle.

1. An example of a low coast—the sandy Algerian beach near Bejaòa.

3. A high rocky coast in Liguria.
4. A high coast in Wales.

2. A low coast in Brittany, scattered with many bivalve mollusk shells.

5. Waves breaking on a coast of hardened lava on the island of Stromboli.

RIVER MOUTHS

1. A satellite photo of the Yangtze River delta, in China. The blue swirls tending toward beige near the coast are deposits of mud and sand. They indicate that the delta is growing. The city of Shanghai appears as a dark patch at the center of the lower half of the photo. It is located on a tributary of the Yangtze. Located in one of the most densely populated areas of the world, with more than 1,000 people per square kilometer (0.386 mi^2), this delta has been radically altered by the creation of artificial channels used to irrigate rice and cotton fields. The channels are the vague, whitish straight lines.

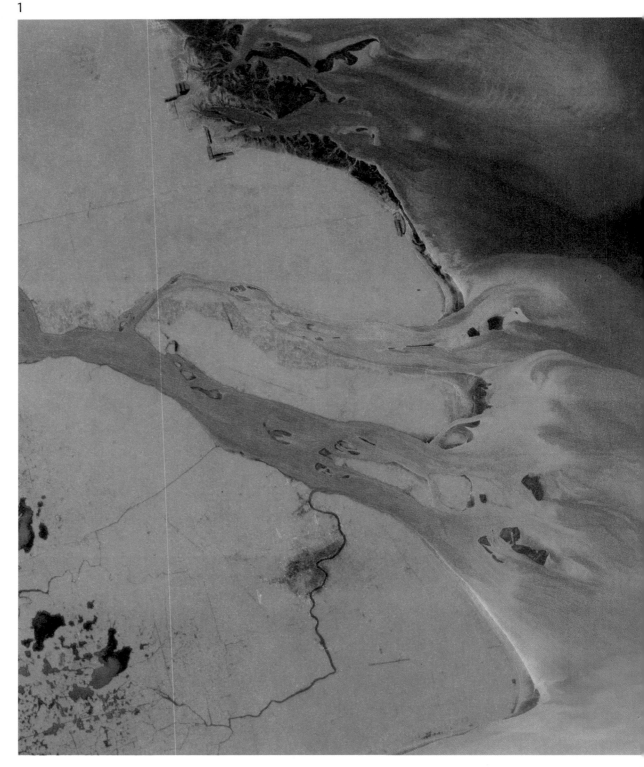

2. The delta of the Nile River in the Mediterranean Sea. Note that the coastline and overall shape resemble a kind of funnel. Along the border, the open seas form lagoons with the barrier islands that are typical of these environments.

3. The "bird-foot" delta of the Mississippi River, in the Gulf of Mexico. In this case, deposition dominates over erosion.

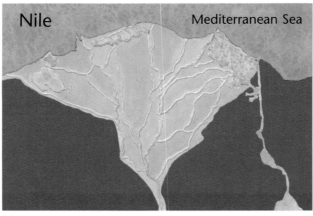

Nile   Mediterranean Sea

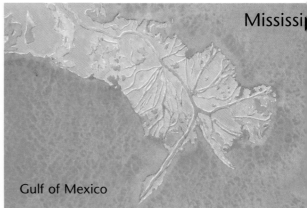

Mississippi

Gulf of Mexico

2   3

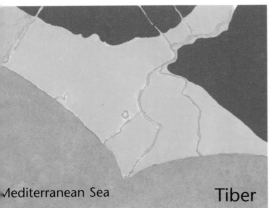

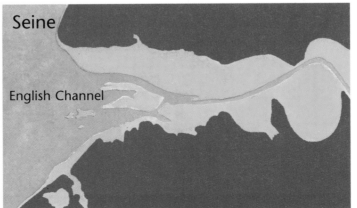

4. A satellite photo of the Rhône River delta, along the French Mediterranean coast. The river appears as a dark line running from the top along with the city of Arles. The river divides into two branches, forming a large triangle, the Carmargue, with a marsh bordered by barrier islands in the center. Other marshes and islands are to the west of the lesser branch. This photo was taken during a flood in January of 1994. The magenta colors show the flooded land. The greens show the woods and fields that were spared by the river.

5. The delta of the Tiber to the southwest of Rome. In this case there is one main stream, but there is still a clearly defined triangular **alluvial deposit**.

6. The delta-estuary of the Seine on the English Channel. The estuary inlet was scoured by tides. It gradually filled with alluvial material, forming a delta. The river then extended its deposits toward the sea. If deposition continues to dominate, the old estuary will gradually be changed into a true delta.

Seine

English Channel

Mediterranean Sea

Tiber

15

Deltas and Estuaries

The buffer zone between the land and the sea is larger and more complex in areas where rivers flow into the sea. Rivers carry large amounts of sandy or muddy matter. This matter is deposited into and then invades the sea. In turn, the sea, through the action of waves and tides, tends to break apart the deposits or build them up into long bars of sand (i.e., sandbars).

muddy expanses. Further inland, estuaries may be bordered by terrestrial vegetation and even woods. This phenomenon is also seen on well-consolidated barrier islands.

River Mouths as Ecotones

A frontier zone between two different ecosystems is called an **ecotone**. An eco-

are often very abundant. In fact, ecotonic environments are the most productive ecosystems on Earth. Within them, photosynthesis may produce as much as 25 grams (0.8 oz) of dry weight per square meter (3.28 ft) each day. Such productivi-

1

The constructive action of rivers tends to dominate on the coasts of enclosed seas. This leads to the formation of large deltas. The deltas are protected from the sea by barrier islands and **brackish** lagoons. In contrast, the erosive action of the sea is dominant on oceanic coasts. Here, rivers end in large inlets called estuaries. At low tide, the muddy floors of the estuaries are exposed. Plants that tolerate salty conditions live along the outer limits of these

tone supports more plants and animals than either of the ecosystems it separates. The **mouth** of a river is an ecotone, since it divides two very different worlds—the terrestrial and the marine ecosystems. To live in this area, where both the water level and the salinity change constantly, organisms must have specific **adaptations**.

Unsurprisingly, estuaries and deltas are populated by only a few species of plants and animals. However, these few species

2

ty is matched only by certain forms of intensive agriculture.

The comparison of the productivity of a natural ecosystem to that of intensive agriculture is illuminating. Such production obviously requires some sort of external energy supply. In the case of cultivated land, this energy is supplied by people and agricultural machinery. In the case of a delta or an **estuary**, however, it is supplied by the activity of rivers and tides. This activity favors the quick spreading of inorganic nutrients and an equally quick removal of waste products.

3

4

1. A view of the estuary of the Conway, in north Wales, at low tide. The vast areas of exposed mud offer food resources to a variety of commuting animals.
2. Low tide in the Seine delta-estuary. The remaining waters flow in natural canals. The canals are seen as a circulation network in the mud.

3. A view of the Fangassier marsh in the Rhône River delta (Carmargue). Brackish lagoons usually form within deltas. The salinity of these lagoons may vary greatly between seasons or even over a few days.
4. A glimpse of Manfredonia marsh. The marsh is a "flood basin," an area that has been artificially surrounded with banks to prevent flooding.

WADING BIRDS

Sand Runners

A large amount of food can be found in the form of flowering plants or algae. This food is used by the vast communities of **invertebrates** that populate deltas, estuaries, and brackish lagoons. These creatures, in turn, support a number of larger animals that prey upon them. One of the most notable groups of larger animals is the birds.

Many birds have adapted to the intertidal zone. During the course of evolution, hundreds of species have developed that rely on methods of feeding such as digging into the wet mud or raking through the material covering the sea floor. Today these birds are classified in about ten different families and are generally known as **waders**.

Waders have fairly long legs, most are good walkers, and some can run quickly. Their ability to swim varies. The plovers, for example, are barely able to swim, but they are good runners. Other groups have webbed feet and are therefore great swimmers, such as the avocets and phalaropes.

Gatherers of Worms

Birds that have adapted to feeding on the muddy plains of low tide search for food either on the surface or by digging. Among the species that search for food on the surface, a favorite tactic is to swoop quickly on small, mobile prey such as crustaceans, small mollusks, and worms. These birds are small, fast, have fairly short legs, and short, robust beaks. Among them are the American golden plover (*Pluvialis dominica*) and the ringed plover (*Charadrius hiaticula*). A more specialized plover, the

1

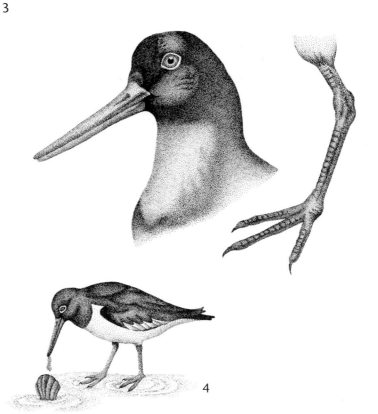

2

3

4

18

sturdy crab plover (*Dromas ardeola*), lives along the coasts of the Arabian peninsula. This bird has a strong beak that allows it to feed on large crabs. The turnstone (*Arenaria interpres*) gets its meals by turning over stones and shells to uncover the small invertebrates that hide beneath them. A similar strategy is used by the wrybill (*Anarhyncus frontalis*), a strange plover from New Zealand. The beak of this bird is bent to the right. It feeds by removing its prey from beneath stones while walking in clockwise circles.

A second group of birds searches for food by digging into the mud. These birds generally have long beaks (ideal for such digging), which they thrust deep into the ground in search of various invertebrates. The group includes many medium-sized or large waders such as the godwits (*Limosa* spp. ["spp." is the abbreviated form of the plural of "species;" "sp." is the singular]) and the sandpipers (*Tringa* spp.).

The mud of the coastal environment contains many bivalve mollusks. Once found, wading birds must open them. The

Photos of wading birds, and detail drawings (not to scale) of their bills, legs, and typical posture on the ground.

1. 2. Common pratincole (*Glareola pratincola*). This bird feeds while flying.

3. 4. Palearctic oystercatcher (*Haematopus ostralegus*). This bird feeds on bivalve mollusks that it opens with its strong beak.

5. 6. Ringed plover (*Charadrius hiaticula*) on a sandy shore.

7. 8. Black-winged stilt (*Himantopus himantopus*). This species can gather small, bottom-dwelling animals in deeper water better than waders with shorter legs.

5

7

6

8

oystercatcher is a beautiful black-and-white wader. It has a long, laterally compressed, red beak and lives on both sandy and rocky coasts. It can efficiently open the mollusks it finds in the mud or detaches from rocks.

Birds With Stilts and Flippers
Some birds gather food by entering the water. These birds risk, step-by-step, finding themselves in water that is deeper than

tom-dwelling organisms. Both genera have long beaks; those of the stilts are straight, and those of the avocets curve upward. Avocets use their curved beaks to stir through the mud while keeping their heads tilted downward. Thanks to their partially webbed feet, these attractive black-and-white waders (white, black, and brown in the case of the American avocet) can swim short distances when they get into water that is too deep for their long legs.

competing for food on the shores by searching for food in different environments. This is the case for pratincoles (*Glareola* spp.). Although they inhabit the coast, the pratincoles have long, swallow-like wings and a large gape that allows them to hunt insects in flight. (This same method is used by swallows, swifts, and nightjars.) They can do this without disrupting other waders.

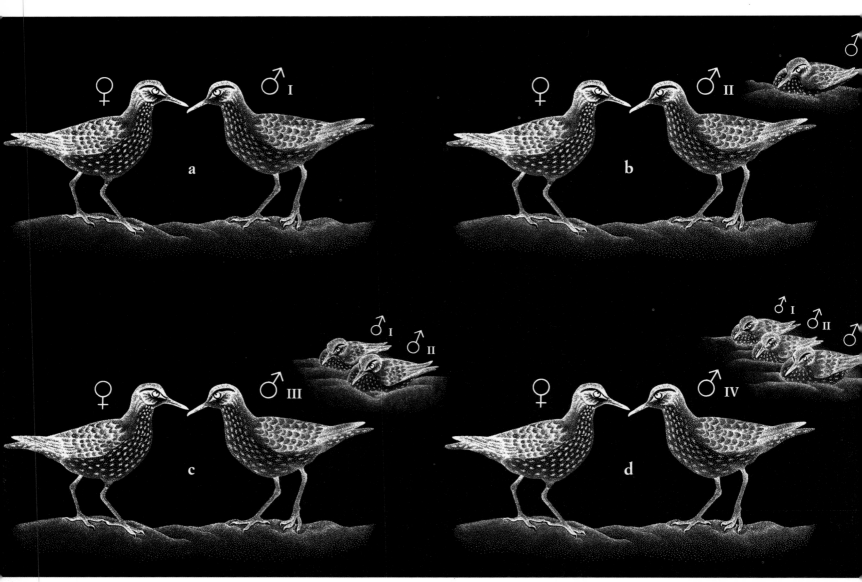

1

they thought. Perhaps for this reason, two genera of waders—the avocets (*Recurvirostra*) and the stilts (*Himantopus*)—have evolved. These birds have very long legs, which obviously are an asset to their method of hunting.

The avocets and stilts can enter water that is 20 to 30 centimeters (7.8 to 11.8 in) deep. Here they can capture various bot-

A further evolutionary step related to feeding in deeper water has been made by the phalaropes (*Phalaropus* spp.). These birds have lobed toes and are good swimmers. They search for food by spinning on the water's surface, creating small vortices. These movements bring nutritious planktonic organisms to the surface.

Some birds have solved the problem of

An Unusual System of Reproduction
Several wader species have adopted a reproductive practice called **polyandry**. This term comes from an ancient Greek word meaning "many husbands." In a typical case of polyandry, the females court the males, mate, and lay a clutch of 45 eggs. After that the males incubate the eggs and raise the young. The females can

then go in search of other males with which to mate. They may repeat the whole reproductive process as many as three or four times during a single mating season. In some cases, the females are larger and more brightly colored than the males (jacanas, for example). In other cases, such as that of the spotted sandpiper (*Actis macularia*), there is no such obvious **sexual dimorphism**.

It may seem like a mystery as to why many of the polyandric bird species are waders. The answer lies in the fact that many waders are migratory. They nest in the marshy areas of the Arctic, Eurasia, and North America and spend their winters on the coasts of temperate or tropical areas. In the far north, the nesting period is brief but very rich in food resources. The laying of many clutches of eggs and raising them at the same time is the best way to make use of these resources.

1. A diagram of the system of reproduction of the spotted sandpiper (*Actis macularia*). This wader belongs to the Scolopacidae family and is common in North America. Following courtship and the fertilizing of the eggs by the first male (a), the female leaves the first clutch, which is incubated by the male. She then accepts the courtship of a second male (b). This male fertilizes the second clutch of eggs and tends to his offspring. In the meantime, the female encounters a third male (c). He takes care of the third clutch of eggs, leaving the female to continue the cycle with a fourth male (d), and so on. In the drawings (b), (c), and (d), you can see the first, second, and third males incubating their respective clutches of eggs. This system is common among wading birds and is known as polyandry (from the Greek word meaning "many husbands"). Its advantage is that it allows intensive use of abundant but short-lived food resources.

2. A classic photo of the wood sandpiper (*Tringa glareola*). This Eurasian wader is closely related to the spotted sandpiper, but it is less likely to practice polyandry.

MANGROVES

Conquering the Mud

In vast areas of the tropics, coastal mud-flats are quite extensive. Found here is an extraordinary group of plants known as the mangroves. Mangroves can colonize the mud and the brackish coastal swamps. They are the central point of an ecosystem that can fight for and obtain new land from the oceans.

The mangroves are very important in Southeast Asia, where violent seasonal rains cause great downward movements of soil, from the forested highlands to the coasts. Because of the shallowness of the sea floor, vast mudflats quickly form and move toward the open sea. For example, the deltas of the Manuk and

As the mangrove community grows toward the ocean, it leaves behind a succession of micro-environments. These range from brackish swamps to dry land by way of freshwater swamps. The mangrove avant-garde is represented by the plants of the *Sonneratia* genus. This group has the highest tolerance for salt water and grows in the area most exposed to the tides. A second group, the *Rhizophora* genus, lives in the strip covered by medium tides.

The most inland group of mangroves are of the *Bruguiera* genus. The roots of these trees are covered by only the highest tides. These tides, known as spring tides, occur twice a month, when the sun and the moon are aligned.

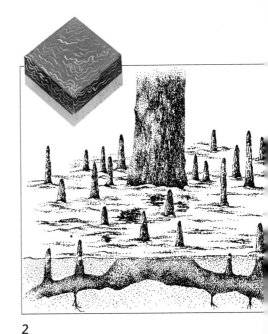

2

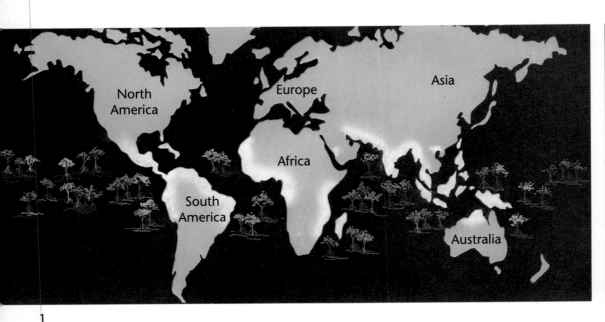

1

3

Solo rivers on Java move toward the ocean by as much as 90 meters (295 ft) in one year. With such a vast area of new land available, the development of a group of organisms that could colonize the inhospitable terrain was inevitable. This is the role of the mangroves. They conquer new mudflats and gradually change them.

Able to tolerate brackish water to some degree (it varies per species), mangroves immediately colonize the mudflats. They combine and connect these flats in an intricate maze formed by their large, aerial roots. In turn, they also gradually raise the level of the land, and in doing so they build a barrier between the ocean and the brackish inland swamps.

The three genera of mangroves are easily distinguished by the formation of their aerial roots. Those of the *Sonneratia* genus branch out horizontally at shallow depths. They often form vertical suckers that look like stakes. The roots of members of the *Rhizophora* genus form a kind of climbing arch. This makes the plants look as though they'd been partially uprooted and then left in the ground. Lastly, the roots of the *Bruguiera* genus grow horizontally. They resemble snakes, with an undulating vertical rather than horizontal movement.

Life in the Kingdom of Mud

The rhythm of life in the mangroves is governed by tidal movements. When the water rises, the more aquatic organisms

4

nter into action. When the tide goes out, it is the turn of the land organisms.

The most successful organisms of the mangroves are those that have best adapted to an amphibious life. Among these are many crabs and an interesting family of fish known as the mudskippers. Mudskippers can leave the water and move across the mudflats like small land animals.

In a mangrove swamp, crabs occupy their usual niche, the intertidal zone. This area is large and highly varied, and the result is a fair number of crab species that have adapted to diverse conditions. Several species, such as *Matuta lunaris* and *Neptunus pelagicus*, can swim, whereas others can walk. A few species of hermit crabs, or pagurians (*Coenobita* spp.), are also present. To protect themselves, they inhabit the abandoned shells of mollusks or land crabs, such as the robber crab (*Birgus latro*).

The most famous crabs of the mangrove swamps are the fiddler crabs of the *Uca* genus. They are called fiddler crabs because of the great difference in size between the male's two pincers. With a little imagination, the larger pincer can resemble a violin being held ready to play. Fiddler crabs exist in East Asia, Africa, and the Americas.

Most fiddler crabs use their giant pincer not for hunting but for defending their ter-

5

1. A map of the worldwide distribution of mangroves. These plants are found throughout the tropics. They perform the same pioneering role everywhere, reclaiming land from the sea.
2. The roots of *Sonneratia*, the genus of mangroves that can grow in the areas most exposed to salt water. Note the elevated level of water in the block diagram and the stake-like breathing roots, or pneumatophores.
3. The roots of *Rhizophora*. These mangroves grow in the intermediate area in the coastal sequence. Note the intermediate level of the water in the block diagram. The climbing-arch structure of the aerial roots is the most common root system in the mangroves.
4. The undulating roots of the *Bruguiera* genus. These mangroves grow in the band farthest from the sea and therefore are the least exposed to salt water.
5. A mangrove of the *Rhizophora* genus on the Thailand peninsula.

ritory, fending off other male crabs, and attracting females. The males continuously click the large pincers to rhythms that vary according to the species and whether they are trying to attract females or repel other males.

American researcher Jocelyn Crane has done a study of fiddler crabs in Panama. She noted that not all females provoke the same reactions among males. The appearance of some females provokes the same reactions given to "threatening" males.

The other group of mud-dwelling animals worth special mention is the afore-

1. Mangroves in tropical Asia. Mangroves are not in just one family in the plant kingdom. They are a group of plants that belong to a variety of families. However, they share a common method of colonizing new terrain in brackish mud.

2. A closeup of the root system of a *Rhizophora* specimen. The mangroves of the Rhizophoraceae family are distributed throughout the world. Those of the Combretaceae, Verbenaceae, and Sonneraticae families are generally limited to Africa, Asia, and Australia.

3. A *Rhizophora* branch with the characteristic hanging, dart-shaped fruit. These fruits contain seeds that often germinate while

1

2

3

mentioned mudskippers. Mudskippers are fish that can emerge from the water and crawl over the mud or climb on aquatic plants. They can do this because of their long bodies and their two pectoral fins, which look like stumpy front limbs with no elbows.

Mudskippers have large protruding eyes. Their sense of sight is very different from that of other fishes and is similar to that of land **vertebrates**. Lastly, their gill covers seal perfectly. This keeps their large

they are still attached to the plant. When the fruits with the germinating seedlings fall, they easily penetrate the soft ground. This is an efficient method of dissemination.

4. A mudskipper (*Periophthalmus* sp.) dragging itself across the mud of a mangrove swamp in tropical Asia. These unusual fish dive into the mud with a speed comparable to frogs.

5. A large *Ocypoda* crab, typically seen running on tropical beaches.

6. A macaque (*Macaca irus*) sitting in the forked branches of a mangrove.

gill chambers filled with water. New oxygen breathed in through the mouth is continually dissolved in this liquid. Thus, the mudskippers breathe via gills like all other fishes, but they have a special system that allows them to breathe when they are on dry land as well.

Mudskippers generally leave the water to hunt for insects. Other fishes, such as the archerfish (*Toxotes jaculator*), also feed on insects, but they hunt while in the water. They capture their prey by shooting jets of water from their mouths. Newborn archerfish have a "spitting range" of only about 10 centimeters (3.9 in). Adults, on the other hand, can shoot down insects that are up to 1.8 meters (5.9 ft) above the water's surface.

The Super-Predators of the Mangroves

Crustaceans and fishes provide a rich food resource to creatures that live in and around the mangrove community. They are eaten by and serve as the main energy source for many creatures. Apart from the otters, of which there are several species in tropical Asia, there are also crab-eating mangustas in Indonesia (*Herpestes urva*), fish-eating cats (*Felis viverrina*) and mongooses (*Cynogale bennetti*), and crab-eating macaques (*Macaca irus*). The macaques are so common and so invasive that they prevent any other monkeys from feeding on crabs even as a secondary food source. The other large Asiatic monkey of the mangroves is the proboscis monkey (*Nasalis lavatus*) of Borneo. The male of the species has a grotesque, 10-centimeter (3.9-in) long nose. Proboscis monkeys are vegetarians. They feed on leaves, flowers, and fruit.

Fishes and crustaceans are also a food source for many birds. Among them are ospreys (*Pandion haliaetus*), fishing owls (*Ketupa zeylonensis*), kites (*Haliaster indus*), skimmers (*Rynchops albicollis*), **cormorants** (*Phalafacrocorax carbo*), darters (*Anhinga rufa*), black-and-white kingfishers (*Ceryle rudis*), herons, storks, and sea eagles.

Two large reptiles of the mangroves of Southeast Asia are considered "super predators." They are the Asian water monitor (*Varanus salvator*), which may grow to 3 meters (9.8 ft) in length, and the estuarine crocodile (*Crocodylus porosus*), which may grow to over 7 meters (22.9 ft). The latter will go not only into the brackish water of the mangroves but also hundreds of kilometers into the open sea. Some have reached the Solomon Islands and even Fiji.

Sadly, mangrove trees have been commercially exploited for years, and the damage this has done to their environment has been extensive. The wood of the mangrove tree has had a multitude of uses, including the building of wharves. Mangrove bark is a wealthy source of various tannins. A tannin is a natural phenolic substance valuable in dyeing, tanning

complex, thus a synthetic version would be impractical, at least from an economic standpoint. The approximate compositional formula for tannin is $C_{76}H_{52}O_{46}$.

True mangroves belong to the division known as the Magnoliophyta. They are in the class Magnoliopsida, the order Rhizophorales, and the family Rhizophoraceae. Other floral species bear the common name "mangrove" but do not in fact belong to the group of "genuine" mangroves.

1. A darter (*Anhinga rufa*) in a tropical Asian swamp. Among the common mangrove animals, this is one of the most unusual. It is an excellent swimmer and often moves through the water with only its head above the surface. It has a long, snake-like neck (it is sometimes known as the "snake-bird" because of this) that can be bent into an S-shape. It quickly straightens its neck when "firing" its sharp bill at its prey.

1

(turning animal skin into leather, since it arrests the decomposition process of certain proteins), the production of ink, and for medicinal purposes (often as an astringent for the treatment of low- to medium-grade burns). Ironically, tannins can be obtained from more abundant trees, including oaks, hemlocks, and chestnuts. Tannin also can be found in coffee and tea. The chemical structure of tannin is highly

2. The head of a Siamese crocodile (*Crocodylus siamensis*), one of the largest predators of the mangroves. It can reach lengths of 3 or 4 meters (9.8 or 13.1 ft), but it is smaller than the Nile crocodile (*Crocodylus niloticus*) or the estuarine crocodile (*Crocodylus porosus*), both of which also can be found in mangroves.

DUGONGS AND MANATEES

Endangered Vegetarians

The mangrove environment is the last refuge of an algae-eating aquatic mammal known as the dugong (*Dugong dugong*). This strange creature is similar to a walrus, but it has no tusks or rear limbs. Dugongs can grow to 3 meters (9.8 ft) in length and can weigh as much as 600 kilograms (1,320 lbs).

The dugong belongs to the order **Sirenia**. This group has obscure origins and questionable relationships with other animals. It also has an uncertain future, and all sirenians are now entered in the *Red Book* of endangered species.

The dugong is a Pacific animal and is in immediate danger of extinction. It shares this status with its closest relatives, the manatees (*Trichechus* spp.). There are three existing manatee species, two American and one African. They are all similar to the dugong in appearance and daily habits.

Dugongs and manatees are basically aquatic. Dugongs exist in coastal regions; manatees along rivers and canals. These are areas in which the border between sea and land is blurred (e.g., the swamps of Florida or the Orinoco basin). These animals feed on sea grass and algae and on aquatic river plants. They tear up the plants with their muscular lips, which are decorated with stiff bristles. Their feeding habits and the mooing-like sound they make has led them to become known as sea cows.

Dugongs swim fairly slowly. They graze under water for up to 10 minutes before surfacing to breathe. Occasionally their head and shoulders rise above the surface in an almost human posture. When they can, they live in family groups. In the most favorable areas, these groups may unite to form small herds.

Groups of dugongs were once fairly numerous. Today they are rare. The females are generally seen with their two youngest offspring. As far as we know, dugongs have no particular reproductive season. During the year, young animals of all ages can be found.

Little is known about the family and the social ties that exist among the various sirenian species. The courtship display of dugongs has been described as "inconspicuous." They mate in water; this goes without saying since sirenians spend their entire lives in the water. They also give birth in water, after a gestation period of about 11 months.

Sirenians appeal to many people. Unfortunately, they also have been a convenient source of meat and fat. As a result, they have been hunted ruthlessly. This situation has been worsened by the superstitions of Chinese medicine. These superstitions attribute miraculous properties to various parts of these animals.

The Sad Story of Steller's Sea Cow

The sirenians' risk of becoming extinct is illustrated by the story of another gigantic representative of the group, Steller's sea cow (*Hydrodamalis stelleri*). Steller's sea cow grew to lengths of 9 meters (29.5 ft) and weighed up to 3.5 tons. It was the only sirenian to conquer the cold seas, and it lived near the islands off Kamchatka, in the northern Pacific. There were never more than a few thousand of them.

Steller's sea cow was discovered in 1741 by sailors and whalers and was hunted to extinction within 20 years. All that remains of the species is the description by the naturalist Steller. The body of these animals was covered with a thick, dark brown, wrinkled skin. The head was very small in relation to the rest of the body. In place of teeth, the jaws had hard, horny plates that the animal used to grind up the sea grass found in various estuaries and along the coasts of the Copper and Bering islands. Steller's sea cow was not a great swimmer. Occasionally specimens were thrown onto the shore by the force of storms.

1. A reconstruction of *Uinatherium*, a mammal that lived during the Eocene. It might be the ancestor of the Proboscidea (which includes modern elephants) and the Hyracoidea (not shown) as well as the sirenians. All these animals appear to be very different at first sight. The Hyracoidea are small and resemble marmots. The ancient ancestry of the *Uinatherium* can be traced only through the analysis of fossil remains. In the case of the sirenians, these are rare and fragmentary.

4. A female West Indian manatee (*Trichechus maenads*) with her offspring. Together with the other two surviving species, the Amazon manatee (*Trichechus inunguis*) and the African manatee (*Trichechus senegalensis*), the West Indian manatees live in the coastal waters and rivers of tropical America and along the coasts of West Africa. These sirenians, which have rounded tail fins, are thought to be more recent than the dugongs from which they may have evolved.

5. A skeleton of *Dusisiren* dating back 19 million years. It is the oldest representative of the Hydrodamalinae, a group of sirenians adapted to the cold waters of the northern Pacific. The area of distribution of this group was gradually reduced over the last 20,000 years. The last species, Steller's sea cow, was hunted to extinction in the mid 1700s.

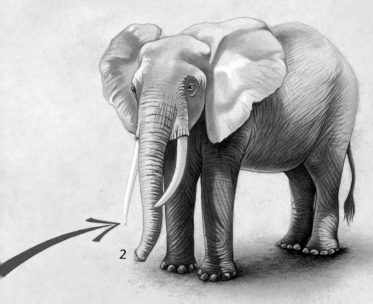

2

2. For a long time, the origins of the sirenians, given the primitive nature of their skulls, were thought to lie with the ancestors of several groups of land mammals. These included the elephants and members of the extinct Embrithopoda order. The Embrithopoda order is represented by the large *Arsinoitherium* of the Egyptian Holocene. According to the American paleontologist Robert Carroll, there is no convincing proof of this descent.

3. The dugong (*Dugong dugong*) is thought to be the oldest of the four surviving sirenian species, even though no fossil remains have ever been found. It lives in the coastal waters of the tropical belt of the Indian and Pacific Oceans. It is distinguished from the manatees by its marine lifestyle and its tail flipper, which is divided in two. In spite of having streamlined bodies, flipper-like appendages, and horizontal tails, sirenians are not particularly good divers or fast swimmers.

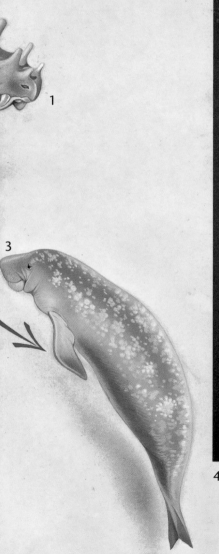

1

3

4

5

An artist's impression of humans hunting Steller's sea cow (*Hydrodamalis stelleri*) in the mid-18th century off the Kamchatka peninsula, in the Bering Sea. The boats carry whalers armed with harpoons. The whalers take aim at the huge but gentle animals as they graze on the soft algae in the shallows of a bay. A silent witness to the impending tragedy, a Steller's sea eagle (*Hailaeœtus pelagicus*) is perched on a rock, patiently awaiting the possibility of scraps.

31

MARINE IGUANAS

In the era when they were colonizing the coastal environments, terrestrial vertebrates evolved into many more carnivorous than **herbivorous** species. Among the mammals, compared with about 100 carnivorous **pinnipeds** (seals, sea lions, and

(*Amblyrhynchus cristatus*).

As immobile as stone statues on the lava rocks of the islands' coasts, marine iguanas are the first animals seen by visitors. Their similarity to miniature dragons (they can be well over 1 meter [3.3 ft] long

1. The head of the marine iguana (*Amblyrhynchus cristatus*), found only on the Galápagos. These reptiles have adapted to an unusual life for lizards—they thrive on a diet of algae and live in a coastal, saltwater habitat.

1

walruses), there are just three existing herbivorous sirenians. In the case of reptiles, **carnivores** are again more numerous. The estuarine crocodile, all sea snakes, and four of the five species of sea turtles (the fifth is **omnivorous**) are carnivores. The only exclusively herbivorous marine reptile is a lizard limited to the Galápagos Islands, this being the marine iguana

and weigh up to 12 kilograms [26.4 lbs]) gives people the impression of being faced with a kind of living fossil.

In reality there is nothing ancient about the marine iguanas; they are actually fairly modern. They have adapted well to conditions on these remote islands, but their ancestors must have looked like the various species of land iguanas found in most

of South America, including the large land iguanas of the genera *Conolophus* and *Iguana*. Marine iguanas are protected by international law.

Marine iguanas are good swimmers, thanks in part to their tails, which are long, strong, and laterally compressed. In the water, they keep all four legs against their bodies and propel themselves only with their tails. The result is an undulating type of movement that is also used by most other lizards while swimming. Marine iguanas can nimbly climb over slippery rocks, and they can dive and remain under water for nearly a half an hour if need be. They get all their food from the sea and eat mainly algae.

If marine iguanas are disturbed while in the water, they will try to return to land. However, if they are disturbed on land, they will move to another spot on land rather than head for the water. This strongly suggests that they do not expect danger on land. Keep in mind that they have been evolving on the Galápagos for many thou-

soft soil is easy to dig. Newborn iguanas must make a brief migration toward the sea, in turn exposing themselves to terrestrial predators (herons, buzzards, **frigate-birds**, etc.). As soon as they hatch, they seek out refuge before taking up their coastal positions, which they often will occupy for most of their lives.

Marine iguanas are not as at home in the sea as the true marine reptiles. In sea water, their body temperature quickly cools from the 36 to 37°C (around 97°F) they achieve by sunbathing, to around 20 to 25°C (68 to 77°F). They must also contend with their main predator, the shark, while in the water.

sands of years, away from all human life, so they don't really recognize us as a threat. As many visitors to the Galápagos can tell you, they hardly acknowledge humans even when approached within a few feet!

Marine iguanas lay their eggs on land, beyond the rocky coastal band, where the

2. A marine iguana set against a map of the islands. A drawing of the *HMS Beagle*, the ship that carried the British naturalist Charles Darwin to the area in 1860, is shown. In his log, the author of the theory of natural selection described these animals in great detail. He was the first to notice that they grazed in algae gardens on the sea floor, relatively far from the coast.

AMPHIBIOUS CARNIVORES

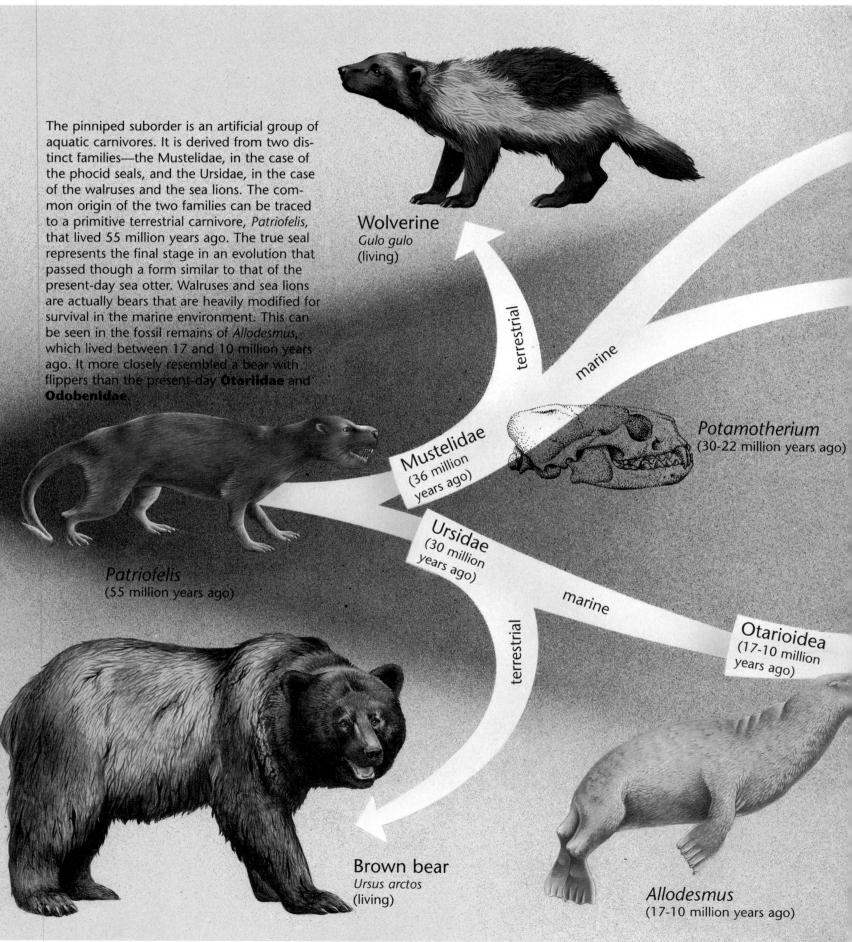

The pinniped suborder is an artificial group of aquatic carnivores. It is derived from two distinct families—the Mustelidae, in the case of the phocid seals, and the Ursidae, in the case of the walruses and the sea lions. The common origin of the two families can be traced to a primitive terrestrial carnivore, *Patriofelis*, that lived 55 million years ago. The true seal represents the final stage in an evolution that passed though a form similar to that of the present-day sea otter. Walruses and sea lions are actually bears that are heavily modified for survival in the marine environment. This can be seen in the fossil remains of *Allodesmus*, which lived between 17 and 10 million years ago. It more closely resembled a bear with flippers than the present-day **Otariidae** and **Odobenidae**.

Wolverine
Gulo gulo
(living)

terrestrial

marine

Mustelidae
(36 million years ago)

Potamotherium
(30-22 million years ago)

Ursidae
(30 million years ago)

marine

Otarioidea
(17-10 million years ago)

terrestrial

Patriofelis
(55 million years ago)

Brown bear
Ursus arctos
(living)

Allodesmus
(17-10 million years ago)

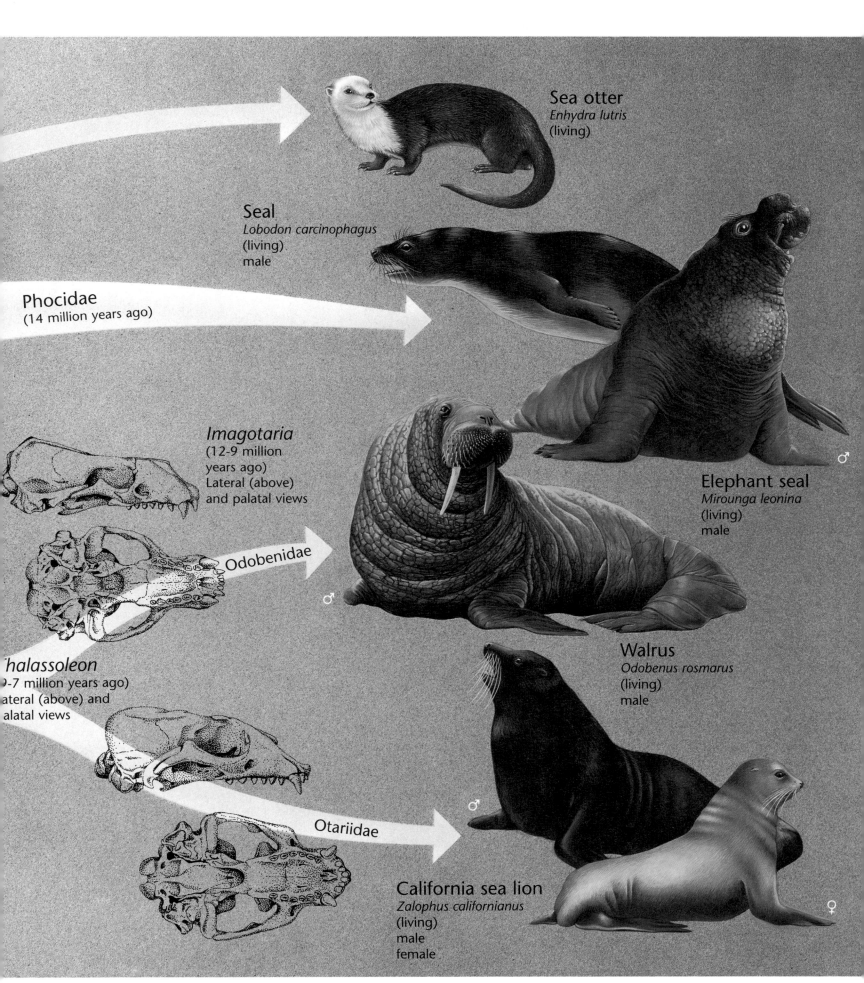

Sea otter
Enhydra lutris
(living)

Seal
Lobodon carcinophagus
(living)
male

Phocidae
(14 million years ago)

Elephant seal
Mirounga leonina
(living)
male

Imagotaria
(12-9 million
years ago)
Lateral (above)
and palatal views

Odobenidae

Walrus
Odobenus rosmarus
(living)
male

halassoleon
-7 million years ago)
ateral (above) and
alatal views

Otariidae

California sea lion
Zalophus californianus
(living)
male
female

35

1. 2. A front flipper (right) and the rear flippers of a sea lion. The digits of the rear flippers are very long and equal in length. The length of the front flippers decreases progressively from the first to the fifth digits. While still retaining nails (visible in the photo below), the digits are connected by a thick swimming membrane.

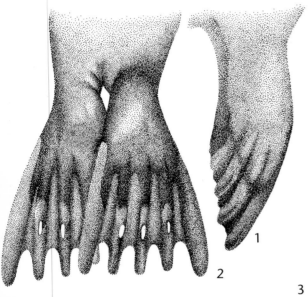

3. The rear flippers of a sea lion. In the case of the pinnipeds, as with other mammals that have adapted to an aquatic lifestyle, evolution has modified their limbs into flippers. The bones of the limbs have shortened and regressed into the chest so that only the "hands" or "feet" protrude. The change in the sea lions' limbs is less extreme. These animals still use their flippers to walk, albeit fairly awkwardly. Seals really cannot walk at all since their limbs act only as flippers.

Perfect Physiological Adaptation

Pinnipeds are a suborder of carnivores that appear to have evolved from ancient animals that were similar to modern bears and weasels. Currently the group is made up of three different morphological types, each considered a separate family. These are the seals, the walruses, and the sea lions. There are about 25 species in all.

Pinnipeds rest and reproduce on land, but spend the rest of their time in the water. This makes them amphibious in the strictest sense of the word (they are, of course, not amphibians).

Pinnipeds have had to face the usual problems of adapting to a marine lifestyle. These problems include absorbing enough oxygen to spend long periods under water, finding fresh water where the only water is salty, avoiding **embolisms** when surfacing, protecting themselves from the cold, and getting enough food to survive and prosper.

Occasionally groups of seals are seen floating vertically in the sea. In that position they can sleep and breathe normally while keeping their heads out of the water. Other species can sleep under water. They periodically surface to breathe without even waking. When they surface, they take in a great deal of air, then submerge again. This concentrated oxygen-intake process is so natural that some species breathe in this way even when sleeping on land. For example, the elephant seal (*Mirounga angustirostris*) breathes slowly for five minutes, then holds its breath for eight, sealing its nostrils. When pinnipeds dive under water, they do not fill their lungs with air like humans. Instead they empty their lungs completely, retaining the oxygen linked to the hemoglobin in their blood.

When diving, the use of oxygen by these animals lessens by as much as 90 to 93 %. Their pulse rate drops from around 80 to about 10 beats per minute, and the flow of blood is concentrated near the brain. This happens due to a system that can close off the arteries that normally carry blood to other parts of the body.

Pinnipeds make full use of anaerobic glycolysis and block the process of cellular respiration. This process takes place in humans and all other vertebrates. What is unusual with the pinnipeds is that they can tolerate concentrations of lactic acid in their bloodstream (oxygen debt) without suffering muscular pain or risking a collapse. Because of these adaptations, pinnipeds can do great underwater feats. For example, they can

dive for periods of up to 40 minutes (varies per species), and many reach depths of 350 meters (1,150 ft) or more.

Problems With Walking

Pinnipeds are generally sociable animals. They often gather on rocky or sandy coasts to rest and bask in the sun, displaying surprisingly human-like behavior. They give birth to and nurse their young on land. They are very cautious on land, remaining close to the water so they can dive in at the first sign of danger.

The limbs of seals—short at the front and curved to form a tail in the back—are essentially useless for walking. On land they must move on their bellies, in a series of jumps. Sea lions and walruses are a bit more versatile. They have rear limbs that can be brought forward, like those of most land mammals. Sea lions are not very heavy and therefore can move fairly quickly on dry land. They are also good rock-climbers. In contrast, walruses are too heavy to move around easily on land.

When they swim, all sea lions and walruses fold their rear limbs back. They use only their front limbs to propel themselves through the water. True seals, on the other hand, swim by moving their rear limbs, which join to form a kind of "tail fin."

Mating and Giving Birth on Land

Although pinnipeds have adapted to life along the coasts rather than in the sea, they feed only on sea creatures (mainly on fishes and crustaceans). Therefore they must limit the time they spend on dry land. The reproductive cycles of the various species take place within a few weeks. During this time, the heavily pregnant females come out of the water, give birth, and nurse their young for a brief period. Then they mate and are fertilized for the following year's cycle.

Some species of pinnipeds have a promiscuous social and sexual system. There are frequent battles among males that are trying to attract females. Other species show **polygyny**. In such cases territorial males attract a harem of females. In the case of the Atlantic gray seal, the harem may include a minimum of four to six individuals. With the hooded seal and the elephant seal, there may be dozens. The males of these last two species are much larger than the females. They also have developed strange-looking appendages on their snouts.

4. A group of sea calves (*Phoca vitulina*) on a northern beach. These carnivores have completely adapted to sea life. They have lost their outer ears (in contrast with the sea lions, which still have vestigial ears) and necks. They cannot move their rear flippers forward because these are folded back to form a tail. Sea cows are common from the pack-ice regions south along the Atlantic coasts of Eurasia and America.

5

5. The head of a submerged sea calf.
6. Galápagos sea lions (*Zalophus wollebacki*). The total population of these animals, the largest on the islands, is about 50,000 individuals. Adult males may weigh as much as 250 kilograms (550 lbs). The Otariidae are distinguished by their coats. Sea lions have rougher coats (*Otaria, Zalophus, Eumetopias, Neophoca,* and *Phocarctos* genera). Those with soft, short coats that were once prized as furs are called fur seals (*Callorhinus* and *Arctocephalus* genera).

6

The Sea Otter

Many mammal researchers believe that today's pinnipeds do not share a common ancestry. Those best adapted to the marine environment, e.g., phocid seals such as the elephant seal, evolved from mustelids similar to present-day otters. The sea lions and walruses, on the other hand, descended from primitive carnivores similar to bears. This theory is supported by the existence of a large mustelid that has an exclusively marine lifestyle, the sea otter (*Enhydra lutris*).

The sea otter may grow to a length of 1.5 meters (4.9 ft) and weigh 40 kilograms (88 lbs). It is found along the coasts of California and western Alaska, as well as near several small islands between Kamchatka and Japan. It inhabits coastal waters, spending the night on floating masses of algae. During the day, it fishes for mollusks, echinoderms, and crustaceans.

Sea otters normally dive for about 90 seconds and reach a maximum depth of 60 meters (196 ft). They do not have the thick layer of fat that protects seals from the cold water. Their only thermal insulation is provided by the air bubbles trapped in their thick fur.

In the recent past, sea otters were hunted for their fur and nearly became extinct. Fortunately they are now protected by environmental law, and their numbers are beginning to increase. Interestingly, they use tools to help them feed. For example, they use stones to detach mollusks from rocks. They often use two stones, one as support and the other as a hammer, to break mollusk shells. Sea otters may also eat fishes and octopuses. In these cases, they use a stone as a kind of "paperweight," holding the remains of their prey in place on their bellies as they float on their backs amid the algae.

A sea otter (*Enhydra lutris*) floating on its back with its rear feet out of the water. This aquatic mustelid is interesting because it is one of the few animals that uses tools. It is also a key species, one that is very important from an ecological point of view. In the past, the hunting of sea otters caused a drastic reduction in their numbers. As a result, the populations of sea urchins on which they fed increased explosively. The urchins, in turn, destroyed the plant populations on which they fed—brown algae or kelp (*Fucus* and *Laminaria*)—and caused the collapse of the entire ecosystem that was based on these plants. A ban on the hunting of sea otters has allowed many populations to begin growing again. This has saved not only the otters, but also the aforementioned ecosystem.

1

BIRDS OF THE ROCKY COASTS

Gulls and Terns

Colonies of seabirds that spend their lives on the sea or along the coastlines and come to land only to reproduce are found on the rocky coasts of all the oceans. In this chapter we will focus on the seabirds that live along the coasts throughout the year. These birds include the gulls, terns, **gannets**, cormorants, frigatebirds, bos' n birds, and **pelicans**.

The most familiar group of coastal seabirds probably is the gulls. There are about 50 species found along all the coasts of the world. In some cases there are large populations, providing amazing examples of adaptation. In many cases, gulls travel far inland via rivers and lakes, feed on rubbish piles, and increase greatly in number. This happens with the black-headed gull (*Larus ridibundus*), the herring gull (*Larus argentatus*), and the ring-billed gull (*Larus delawarensis*). Other gulls have a very limited geographical distribution or are closely associated with the marine habitat. An example of the first case is *Larus audouini*. This species is present only in the Mediterranean in a few thousand pairs. Two good examples of the second case are the black-legged kittiwake (*Rissa tridactyla*) and the red-legged kittiwake (*Rissa brevirostris*). They normally live out at sea and come to land only to nest.

A group that is fairly closely related to the gulls is the terns. These birds have forked tails and longer and more pointed wings than the gulls. They also are more agile fliers and are expert fishers. Terns can be found in all the oceans. The Arctic tern is an example of a long-distance migratory bird. It nests in the Arctic and spends the winter in the Antarctic.

Specialized Fishers and Pirates of the Air

Gulls and terns fish in shallow waters. Cormorants, gannets, and frigatebirds are more specialized and do their fishing farther out at sea. The most common of these birds are the cormorants (*Phalacrocorax* spp.).

There are about 40 cormorant species distributed around the world. Some favor marine environments, others freshwater. Cormorants swim well and are able to adjust their buoyancy, staying close to the

surface like ducks or semi-submerged like darters. Although they are good fliers, they also dive and swim well. However, their wing feathers are not waterproof and therefore quickly become soaked. This forces the cormorants to spend a lot of time drying their wings. They often stand with their wings open for this very purpose.

Gannets are more dramatic than cormorants in their hunting habits. They plunge into the water from high above the surface. Like cormorants, they nest in colonies, and all nine species favor marine habitats. They spend from six to eight months each year in crowded colonies on remote coasts or small, rocky islands, and their nests are spaced well apart for safety reasons. The most common species is the Atlantic gannet (*Sula bassana*). Their entire population reproduces in just 22 colonies, 14 of which nest in the British Isles.

The large and spectacular frigatebirds (*Fregata* spp.) are well known in various parts of the oceans. There are five species, and all are darkly colored. Frigatebirds have long, hooked bills, very long, curved wings, and deeply forked tails. During their reproductive period, the males are adorned with large red throat pouches, which they inflate when courting the females. Frigatebirds nest in colonies on small islands along the coast. They feed while flying by grabbing fish that jump out of the water. They also persecute other birds, forcing them to regurgitate the prey they have already swallowed.

1. A sheer, rocky cliff overlooking the sea in northern Europe and populated by several species of nesting seabirds.
2. The black-legged kittiwake (*Rissa tridactyla*) in flight. In contrast with the black-headed gull and the herring gull, the kittiwake lives only at sea except during the reproductive cycle.
3. A closeup of a rocky cliff with the nest of a cormorant (*Phalacrocorax* sp.). This animal feeds under water but generally stays near the coast.
4. A group of terns (*Sterna* sp.) on a beach. These birds nest in low and easily accessible sites, often on beaches or grassy cliffs.

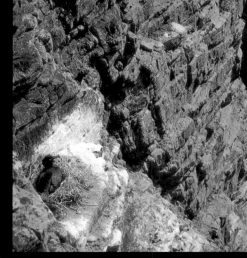

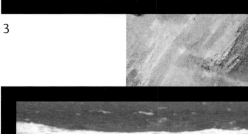

2

1. 3. A male frigatebird in the Galápagos with its throat pouch inflated and deflated. The inflation of the throat pouch is the main way of attracting females.

2. Blue-footed boobies (*Sula nebouxii*) on the Galápagos Islands. The gannet family includes nine species, six of which live in tropical oceans.

4. A group of lesser black-backed gulls (*Larus fuscus*). This is a slightly darker northern version of the herring gull. It feeds on refuse and is also a predator of other birds. **5.** A gannet (*Sula bassana*) sitting on its nest. Among the Sulidae family, this is the most familiar species. It is the only one found on the Atlantic coasts of Europe and North America. After a ban on hunting was imposed at the end of the last century, its numbers began to increase. The population is still growing.

An arctic tern (*Sterna paradis-* with a small fish in its bill. bird is the greatest living rant—each year it flies from Arctic to the Antarctic.

A colony of gannets on the ks of the island of Bass, in tland, which gave its name he species (*Sula bassana*).

THE COASTS AND HUMANS

Coastlines are unstable environments. They are subject to continual change because of tides and rivers, changes in climate, and changes in sea level. For economic reasons, coastlines are also exposed to various human activities. The coasts may be subjected to land reclamation, settlement, fires, and vandalism.

Land Reclamation

Along the coasts, the actions of the seas and rivers often gives birth to lagoons and swamps. These are among the most productive environments from an ecological point of view. They are also among the most attractive. For centuries, humans have attacked these environments with land reclamation projects. In these projects, the water is drained from the swamps in an attempt to change the expanses of mud or reed beds, in which herons and avocets nest, into cultivated fields. Only a small fraction of the coastal wetlands that existed in the last century have survived, even near the great rivers. The wild populations of plants and animals that lived in these environments have also been reduced.

Urban Settlements

The coasts are a favorite environment of humans. Apart from estuaries and deltas—areas favored because of their freshwater supply and easy access to rivers and seas—all sheltered bays are potential targets for the establishment of villages as long as a local supply of fresh water is available. Thus, coastal villages are found throughout the world. Some are very picturesque, and their economies are closely tied to the sea.

Apart from these villages, people have also erected industrial complexes and tourist centers. As a result, many a coastline has changed into a linear megalopolis, and the natural environments have been so compacted that they become nearly invisible. One disturbing aspect of this situation is that the average person is now able to conceive only a coastline that consists of theme parks, casinos, restaurants, and other such attractions.

A dead sea duck covered with crude oil that was spilled by a tanker. Accidents like this cause the death of many seabirds.

A classic image of a marine disaster—an oil slick and a fire that occurred on June 23, 1987 following a collision between the Greek tanker *Vitoria* and the Japanese tanker *Fuyoh-Maru,* near Aiziers in Normandy. Dramatic accidents such as these oblige us to ask how much we are prepared to pay for continued technological development.

Istanbul, 1988. This image of ecological degradation speaks for itself.

Roads

Roads and railways have also changed the coastlines. Even when alternate roads and railways are available, most travelers follow the coastal routes because they provide the easiest way to connect with other coastal settlements. Some roads follow the coastline for purely economic reasons. In fact, roads are often built before any settlements. They are then used to aid in the building of houses, commercial structures, and so on. At the same time, the delicate balance of the natural environment is broken apart. Maintaining natural environments is the only situation that can guarantee the conservation of biological diversity. However, it is destined to collapse in the face of increasing human occupation.

Trash accumulated on the beach at Rarotonga, in the Cook Islands.

Pollution

Many people believe that urban settlements are responsible for much of the world's pollution. Industrial complexes are, to varying degrees, responsible for a lot of chemical pollution. Example—a major cause of coastal pollution is oil spills. Ships and boats also cause pollution through the residues of the fuels they use and the trash that their occupants throw into the sea. These materials often wash up on the shores and beaches.

Some of the most remote and isolated beaches are in terrible condition because they do not benefit from cleaning services. Trash is usually only an esthetic problem for humans, but for wildlife, it often is toxic and deadly. For example, many sea turtles have perished after mistaking plastic bags for tasty jellyfish.

Firefighters at work on the still-smoldering ashes of one of the many fires that plague the coastal forests of the Mediterranean each year. These fires destroy great expanses of maquis (thick and scrubby underbrush), and the work of the firefighters continues long after the flames are extinguished.

Fires

Particularly in regions that have a very dry season, fires devastate vast areas of the natural environment. However, it has recently been shown that these often are less disastrous than one might think. For example, many natural environments of the Mediterranean have vegetation that can quickly regenerate after a fire. Still, there are many other faunal species that don't re-establish themselves as quickly. Moreover, it is not uncommon for a local community to unwisely decide to proceed with a program of development in an area left barren by a fire.

48

Coastal areas are often chosen as sites for industrial complexes and therefore are subject to water and air pollution. This photograph, taken at the Berre Marsh in France, illustrates a dramatic example of air pollution.

GLOSSARY

Words in *italics* can be found elsewhere in the glossary.

adaptation Term used to describe the adjustment of the traits of an organism in order to better survive in its environment. Long-term adaptation is genetic and occurs in a population through natural selection. Short-term changes in an individual may occur through modifications in its physiological responses or as behavioral changes through learning.

alluvial deposit Debris carried by a river and dropped at points where the speed of the flow decreases. Such *deposition* may form large *delta*s or alluvial plains, such as the Pianura Padana, in Italy.

brackish Term used to describe water that is neither fresh nor *marine*, but rather a mixture of both.

carnivore Term applied to an animal that feeds only on other animals.

cormorants Family of medium-to-large aquatic birds that are generally dark in color and sometimes have a metallic green or blue sheen. Cormorants also have long, slim bills that are hooked at the tip, webbed feet, fairly short wings, and a long tail. There are around 30 species throughout the world.

delta Roughly triangular-shaped geographical feature at the *mouth* of a river, caused by the *deposition* of *sediment*.

deposition Process in which *sediment* carried by wind, moving water, or ice is dropped or deposited in an area.

ecotone A generally narrow and often well-defined transitional zone between two different ecosystems (e.g., between a wood and a meadow) and characterized by a separate community. One of the most common ecotones is the one between dry land and fresh or salt water. Artificial ecotones caused by the cultivation of land are common.

embolism A sudden interruption of the circulation of blood due to the formation of solid, liquid, or gaseous substances, such as an air bubble within the blood vessels.

erosion The process in which *sediments* are carried away from an area by wind, moving water, or ice.

estuary Coastal body of water in the form of an inlet of the sea in which fresh and salt waters meet and mix, forming *brackish* water. Estuaries normally form at the *mouths* of rivers that flow into open seas and have fairly pronounced *tides*.

frigatebirds Family of large birds of the Pelecaniformes order that have long, narrow wings, slim, elongated bodies, mostly black plumage, and hooked bills. During the courting period, the males inflate the red pouches on their throats. Frigatebirds are excellent fliers but cannot swim and do not have waterproof feathers.

gannets Family of large, fish-eating seabirds. Gannets have webbed feet and long, robust, conical bills that lack nostrils. These birds inhale the air they breath exclusively through their mouths.

herbivorous Term applied to animals that feed only on plants or plant products.

invertebrates Animals that lack a backbone.

marine Relating to the oceans or seas.

mouth Terminal, or end, part of a river where it enters the sea or a lake or where it joins another river. It may take the form of a *delta* through *alluvial deposits*, where the river flows into an enclosed sea, or an *estuary*, where an inlet is formed by the actions of the waters of the river and the open sea.

Odobenidae Family of *pinnipeds* related to the *Otariidae*. The family consists of a single species—the walrus—which lacks external ears and has rear limbs that can be moved forward to help the animal move on dry land. In both sexes the upper canine teeth form large tusks.

omnivorous Relating to animals that feed on both plant and animal matter.

Otariidae Family of *pinnipeds* that have small external ears and rear limbs that can be folded forward to aid with movement on dry land. The Otariidae descended from the ancient Ursidae.

pelicans Family of large aquatic birds with long, broad wings, webbed feet, and large beaks that have expandable pouches on the lower jaw. Pelicans live in coastal areas, especially estuaries, feeding on fishes and crustaceans.

Phocidae Family of *pinnipeds* lacking external ears and having rear limbs that cannot be folded forward to help the animal move on dry land. Phocidae are found in all the oceans, mainly in high latitudes, and are better adapted to the *marine* environment than the *Otariidae*, with whom they have no common ancestry. The Phocidae descended from the ancient Mustelidae.

pinnipeds Carnivorous mammals with amphibious habits that have adapted to the *marine* environment. Pinnipeds have streamlined bodies, small heads, nostrils that can be closed when the animal is diving, and limbs modified into flippers with the digits connected by a swimming membrane. There are three families—*Phocidae*, the *Odobenidae*, and the *Otariidae*.

polyandry Reproductive practice in which a female mates with more than one male, either to increase the percentage of fertilized eggs (called simultaneous polyandry—common among insects and other *invertebrates*), or mates with one male, delegating the task of rearing the young to him so she can mate with another (called successive polyandry—common among many wading birds).

polygyny Reproductive practice in which a male normally mates with more than one female but plays no part in rearing the offspring.

sediment Material deriving from the fragmentation of rock and the shells and skeletons of dead organisms. Sediments are gradually deposited in layers on the bottom of the sea or on the surface of the land.

sexual dimorphism Presence of one or more different external features that distinguish the males and females of a species. For example, the males may be larger or more brightly colored. Sexual dimorphism may be genetically based, or it may be hormonal.

Sirenia Order of *herbivorous* mammals totally adapted to the *marine* environment. They have front limbs with five toes that are linked together but still fairly mobile, no rear limbs, rudimentary pelvis-

es, and tails modified to form a horizontal tail fin. Young are born at sea and are nursed via mammary glands.

terrestrial Term referring to the land. Terrestrial organisms are those that typically live on land.

tides Periodic movement of the waters of the seas and oceans caused mainly by the gravitational pull of the moon and, to a lesser extent, of the sun.

vertebrates Animals that have a backbone.

waders Generic term used to describe the birds belonging to the Charadriidae, Glareolidae, Scolopacidae, Haematopodidae, and Recurvirostridae families, in the Charadriiformes order. They are birds that generally have long legs adapted to walking on land or in shallow waters of sandy or muddy coasts.

FURTHER READING

Bramwell, Martyn. *The Oceans* (revised edition). Watts, 1994

Brown, Vinson. *Exploring Pacific Coast Tide Pools.* Naturegraph, 1996

Crump, Donald J., ed. *The World's Wild Shores.* (Special Publications Series 24: No. 4). National Geographic, 1990

Editors, Raintree Steck-Vaughn. *The Raintree Steck-Vaughn Illustrated Science Encyclopedia* (1997 edition). (24 volumes). Raintree Steck-Vaughn, 1997

Farmer, Wesley M. *Seashore Discoveries.* W. M. Farmer, 1986

Hecht, Jeff. *Shifting Shores, Rising Seas, Retreating Coastlines.* Simon and Schuster, 1990

Hester, Nigel. *The Living Seashore.* (Watching Nature Series). Watts, 1992

Howard, Jean G. *Bound by the Sea: A Summer Diary.* Tidal, 1986

Kerrod, Robin. *Birds: Waterbirds.* (Encyclopedia of the Animal World Series). Facts on File, 1989

Lambert, David. *The Pacific Ocean.* (Seas and Oceans Series). Raintree Steck-Vaughn, 1996

Lambert, David and McConnell, Anita. *Seas and Oceans.* (World of Science Series). Facts on File, 1985

Mara, W. P. *Iguanas.* Capstone, 1996

Mattson, Robert A. *The Living Ocean.* (Living World Series). Enslow, 1991

Miller, Christina G. and Berry, Louise A. *Coastal Rescue: Preserving Our Seashores.* Simon and Schuster, 1989

Neal, Philip. *The Oceans.* (Conservation 2000 Series). Trafalgar, 1993

Parker, Steve. *Seashore.* Parkwest, 1992

Quinn, John R. *One Square Mile of the Atlantic Coast.* (One Square Mile Series). Walker, 1995

Stone, L. *Seashores.* (Ecozones Series). Rourke, 1989

Swenson, Allan. *Secrets of a Seashore.* (Secrets of... Series). G. Gannett, 1981

Waterlow, Julia. *The Atlantic Ocean.* (Seas and Oceans Series). Raintree Steck-Vaughn, 1996

Staff, Wildlife Education, Ltd. *Sea Birds.* (Zoobooks Series). Wildlife Education, 1992

PICTURE CREDITS

INDEX

Note: pages in *italics* indicate illustrations